SpringerBriefs in Climate Studies

For further volumes:
http://www.springer.com/series/11581

Krishna Rao Pinninti

Climate Change Loss and Damage

Economic and Legal Foundations

 Springer

Krishna Rao Pinninti
Princeton, NJ
USA

ISSN 2213-784X   ISSN 2213-7858 (electronic)
ISBN 978-3-642-39563-5 ISBN 978-3-642-39564-2 (eBook)
DOI 10.1007/978-3-642-39564-2
Springer Heidelberg New York Dordrecht London

Library of Congress Control Number: 2013943980

Printed on acid-free paper

Springer is part of Springer Science+Business Media (www.springer.com)

Preface

This Special Monograph is prepared with the purpose of laying down a few major foundations and guidelines in the current process of devising strategies, mechanisms, and providing resources for containing or mitigating loss and damage arising from sustained adverse impacts of climate change and climate variability. Whereas the UNFCCC and other entities have been preparing to offer groundwork to facilitate the processes at the international level, it is relevant to augment these efforts with additional insights. This is the attempt of this publication, with an approach of pragmatism and an objective specification of relevant frameworks for further actions.

Development of integrated approaches to the assessment and reduction of loss and damage due to climate change (including climate variability), encompassing economic, and legal dimensions is the primary objective for this Monograph. Assessment of costs (in all their dimensions) of climate change is not an objective here. Similarly, review of literature is not a part of coverage here. This publication is not expected to form a Handbook or Toolkit; this is aimed at top-level policymaking and strategy development, both in the national and international contexts. The readership includes policy analysts, researchers, and top-level policymakers. Actionable knowledge goes beyond provision of information and is thus a global public good. The development of this Monograph takes this route for further development at all levels in the interests of larger humanity.

Princeton, NJ, USA, May 17, 2013   Krishna Rao Pinninti

Contents

Chapter 1
Climate Change Governance

1.1 Introduction

Climate change (CC), with all its ramifications and complex unprecedented adverse impacts, is now seen as a major contributor to natural hazards and disasters around the planet. The scale of these effects of CC- including frequency of occurrence of events- has been of exceeding magnitudes relative to the coping capacities of most developing countries. This ongoing and emerging situation warrants consistent and scaled up efforts on all fronts and by all entities, individually and collectively, to minimize the adverse effects lest we could face collective irreversible losses in almost all spheres of life on this planet. This reading need not and should not imply misdirecting scarce resources toward hypothetical problems and or their fictional solutions. At national and international levels, vision and commitment is important in the design and implement of relevant policies for implementation in this context. This calls for an improved understanding of the climate change systems and their governance, as well as the impacts on the socio-economic and other systems.

Natural hazards and extreme events are on the rise (IPCC 2012). These can result in excessive losses and damages if disaster preparedness is limited. Disaster results from a combination of the following simultaneously operating streams of exogenous and endogenous shocks to physical, economic, and social systems (Rao 2013): (a) financial crises enhance the exposure of the socio-economic vulnerable segments of societies; (b) climate change adversely impacts societies and physical systems with extensive and intensive extreme events; and, (c) precisely at a time when the vulnerable populations need enhanced support do most governments tend to reduce support, with little provision of offsetting mechanisms such as risk insurance or infrastructure development to create economic expansion and jobs or income generation.

Financial crises and budget deficits have resulted in a series of fiscal measures with so-called austerity measures in several countries. There have been new limitations on financial resources that are available for some of the social safety

K. R. Pinninti, *Climate Change Loss and Damage*, SpringerBriefs in Climate Studies, DOI: 10.1007/978-3-642-39564-2_1, © The Author(s) 2014

nets and other programs important for protecting the socio-economic vulnerable sections of the affected countries.

At the outset it is useful to note that *an extreme event or slow onset adverse impact need not turn into a disaster; hazards equate to disasters when the relevant infrastructure (physical, technological and human) lags behind in its capacity to address the needs, be it because of the scale of the adverse event or its timing and repetitiveness or occurrence of multiple hazards—sequentially or simultaneously.* The role of governance in this framework remains among the most important elements, along with vision and commitment to take advantage of modern technologies and other resources to prevent, manage and mitigate the adverse effects of climate change, apart from preventing climate change itself in the first place.

1.2 Mitigation, Adaptation, Disaster Reduction and Effective Governance

Climate change governance (CCG) comprises a set of mechanisms—including institutions, policies and programs- that influence the processes affecting and affected by climate change (Rao 2011). This comprises: (a) Climate Change Mitigation (CCM) via reduction of emissions of greenhouse gases (GHGs) and other mechanisms such as carbon sequestration; (b) Climate Change Adaptation (CCA) that caters to various adjustment mechanisms at all levels needed to seek resilience of systems and their functioning to the adverse effects of CC; and, (c) institutional mechanisms, beyond ad hoc approaches, to prevent and remedy adverse effects of CC at international, national and sub-national levels and across sectors; these include legislative arrangements for coordinated actions, legal liability and compensation mechanisms for damages arising from CC or its contributory ingredients, capacity building, adaptive learning and reform activities. Generally, it is desirable that efficiency (financial, economic, technical and institutional), equity (across regions, countries and sections of society—especially vulnerable areas and people), and effectiveness (meeting the desirable and stated goals and objectives) of various activities is gauged in the design and implementation of relevant policies in any time horizon, scale of operations and their governance. Ideally, formal analytical solutions may be derivable using dynamic multi-criterion optimization under flexibility and uncertainty, with the objectives of maximizing sustainable development at all scales and areas. Several approximations are attempted in the policy and international mechanisms areas, some of which have led to international agreements. These are generally based on group coalitions and strategic considerations as perceived by individual countries, and tend to fall short of paying full attention to the imperatives of comprehensive approaches; synchronization and effectiveness of policies in tune with the required achievement of objectives remains imperfect.

CCA has come to the forefronts as one of the main aspects of CCG relatively recently with increasing realization that: (a) we are falling behind on drastic reduction of GHGs; and, (b) the existing concentration of these causes seems to accelerate the pace of CC and its adverse impacts. Accordingly, significant additional resources need to be made available in order to address the CCA aspects, both in developing as well as developed countries. This makes it imperative that governance issues must be addressed soon in order that the contemplated policies and their governing institutions are up to the task. There is an urgent need for building adaptive capacity at all levels and sectors in most systems. Similarly, the role of multiple approaches needs to be fully examined in order to take advantage of benefits of adoption of such mechanisms and norms of governance. Also, *the governance issue becomes a high priority in light of large-scale linkages and integration requirements of adaptation with the rest of the economic systems and with disaster risk reduction (DRR).* This remains an area of unprecedented complexity: institutional, financial, economic and legal.

International databases, and country profiles of disasters are available from the Centre for Research on the Epidemiology of Disasters (CRED) and related websites: www.cred.be www.emadat.be, www.emdat.be/result-country-profile Let us recall one of the standard definitions of 'disasters'. CRED defines a disaster as "a situation or event which overwhelms local capacity, necessitating a request to a national or international level for external assistance; an unforeseen and often sudden event that causes great damage, destruction and human suffering". For a disaster to be entered into the database, at least one of the following criteria must be fulfilled:

- 10 or more people reported killed;
- 100 or more people reported affected;
- declaration of a state of emergency;
- call for international assistance.

1.3 Close Proximity of Adaptation and Disaster Reduction

Despite considerable understanding of role of effective adaptation in reducing the effects of disasters or of extreme and slow onset events, current CCA measures are usually incremental in their scope and scale; these are responses to CC and are extensions of actions and behaviors that sometimes reduce the losses or enhance the benefits of natural variations in climate and extreme events. These are usually considered under various categories of CCA measures around the world, and can be classified as 'responsive adaptation' measures. In another but not entirely exclusive classification, 'pro-active adaptation' becomes relevant when we can foresee the future needs and the cost-effectiveness and/or implementation feasibility of adopting CCA measures in a progressive manner over time. Adaptation deficit correlates directly with potential loss and damage (L and D) due to various hazards. Risk reduction is better facilitated with greater preparedness, infrastructure development—including adaptive capacity of people and institutions.

Encompassing these, there are at least three classes of adaptations that are considered as transformational (see also Kates et al. 2012): those that are adopted at a much larger scale or intensity, those that are truly new to a particular region or resource system—including innovations in technology and technological adoption, and those that transform local areas such as shifting residences away from being too close to the sea coast. Some of these are collective adaptations that would be explicitly planned and implemented, but they also include autonomous adaptations by individuals and organizations that can cumulate in transformative adaptations, or actions intended to address other problems that can become transformative CCA. Here comes the close proximity to the intersecting areas of DRR and CCA: first order approximations to either streams of activity may seem separable from the other but higher order approximations that take longer time horizons and scope/scales of activity make it rather obvious that we cannot be effective in either CCA or DRR without recognizing the close dependency of each of the two, especially in terms of resources and governance mechanisms that link the two. This is not to suggest that CCM is largely excluded in this discussion: some of the measures in CCM such as reduced deforestation is also an activity that allows CCA and caters to DRR in some areas such reducing mud slides.

Accordingly, the inter-relationships among various elements of CCG need to be understood in any mechanism that seeks to cater to one or more of the objectives in the international and other contexts. An understanding of these interdependencies enables possible assignment of relative responsibilities of nations: those that are affected adversely, and those primarily responsible or have resources to enable the climate change victims to remain resilient and possibly gear up themselves for new scenarios of effects with least loss and damage. Being vulnerable is a relevant criterion for provision of compensation for losses and damages; even a good degree of compensation of may not be sufficient to ensure resiliency at national and sub-national levels. The roles of other inputs, institutional quality and governance provide sustainable infrastructure in effecting CCA and addressing adverse events.

These transformative adaptations, like incremental in adaptation, can be responsive or pro-active/anticipatory. The former take place during and after serious climate change impacts; anticipatory actions occur in advance of threats that pose serious risks of significant impacts. Three conditions set the stage for the application of transformational CCA, in close coordination with DRR measures: (a) large vulnerability in certain regions, populations, or resource systems; (b) sustained disturbances in weather over time in any given region or affecting specific sectors systemically; and, (c) severe CC that threatens to overwhelm even robust human–environment systems. In this context it is also important to note some of the shortcomings of CCA approaches in general. These arise from the prevalence of the phenomenon of maladaptation, explained below.

1.3.1 Maladaptation

Maladaptation is the result of inefficient choices of strategies and policies in any category of adaptation measures that eventually contribute to worsening of the adaptation potential over time and/or scale or in combination with other interventions. It is also sometimes viewed as the result of the adaptive responses made for various interdependent systems without due consideration of adverse impacts on other components (that may or may not be sufficiently influenced by climate change). Externalities of adaptation measures can lead to maladaptation, and it may be more meaningful to restate as follows:

> maladaptation is a phenomenon in the category of externalities that needs to be constantly and fully taken into account in various aspects of the governance of CCA.

In this sense, the IPCC definition (given below, source www.ipcc.ch/pdf/glossary/tar-ipcc-terms-en.pdf) requires modification on the above lines in order to incorporate the role of externalities as the driving principle in maladaptation:

> any changes in natural or human systems that inadvertently increase vulnerability to climatic stimuli; an adaptation that does not succeed in reducing vulnerability but increases it instead.

It is important to note that maladaptation phenomenon prevails as a rule, rather than as an exception, if adaptation policies are designed for a sector without due consideration of its linkages (forward and backward) with other related sectors of the system under consideration (Rao 2012). We need the concept of 'adaptive capacity' when examining loss and damage due to CC. This is explained below. These aspects become very relevant when examining the scope for and operationalization of mechanisms to address loss and damage, as see later in this monograph, when the dependencies of disasters impacts are seen in terms of factors outside the CC system.

Adaptive capacity (AC) is the ability of a system to adjust to climate change and its adverse impacts to cope with the consequences or take advantage of opportunities (IPCC 2007).

CCA includes AC as one of the measures. Whether or not AC is qualified by the requirements of a combination of efficiency-equity-effectiveness criteria depends on the specifics of the components of the AC system and its relationship with the rest of socio-economic, physical and other systems of an area or society.

AC and autonomous adaptation including innovation require more attention than has been accorded so far in any system, since these directions offer relatively more cost effective mechanisms of attuning toward CCA and DRR. Similarly, various traditional assumptions about future scenarios and implications for adaptation need critical scrutiny, with the roles of thresholds and non-linearities in the effects of greater CC (as in a 4 °C change in global mean temperature in contrast to 2° change largely explored so far). The roles of transformational CCA and of large scale activities under DRR become pertinent in some of the potential scenarios of worsening temperature rises and of non-linear effects of such changes on adverse impacts of CC.

Finally, there is no valid reason to treat CCM, CCA or DRR as stand-alone activities. These can easily be mainstreamed into various development activities and dovetailed to meet the specified and desirable objectives and targets. Governance issues need to pay special attention to the CCG activities but not a lot more than what is required in resource allocation and utilization in an effective governance sense for the socio-economic system itself. Specialized and specific interventions in CCG require appropriate attention in governance in an innovative not bureaucratically defined way—whether at the levels of international entities or national/sub-national levels. The formation of relevant capacities in tune with the imperatives of new challenges posed by CC and its complex adverse impacts belongs in the areas of high priority attention, and far beyond current pace of progress.

Effective governance remains possibly the most cost-effective mechanism toward the attainment of the aspirations of the societies, be it to cater to the extensity and intensity of hazards or in attaining the objectives of sustainable development (SD), or both. Governance issues are of paramount significance at all levels, international, national and sub-national. Loss and damage accruals over time are strongly influenced by such factors and these need to be addressed within the framework of devising compensation mechanisms for loss and damage. It would be naïve to leap to some market mechanisms without full assessment of the efficacy of such institutional arrangements. The emissions trading systems have at best resulted in mixed results, and founded on thin markets with little room for (a) market efficiency, and (b) provision of dynamic incentives for polluters to adopt technical and other innovations. These experiences are fairly indicative of the serious limitations of several of the UN Framework Convention for Climate Change (UNFCCC)-supported approaches and warrant a fresh analysis that starts with international duties and obligations for compensation due to climate change and adverse impacts. The Montreal Protocol to the Vienna Convention for Depleting Ozone remains one of the most successful treaties that resulted in desired results, and this did not avail so-called market mechanisms being attempted currently.

1.4 Integrative Mechanisms for CCA and DRR

The recent assessments of the 2012 Special Report of the IPCC *Managing the Role of Extreme Events and Disasters for Advancing Adaptation to Climate Change* (known as SREX Report) deserve high priority attention. The processes CCA, DRR, and Disaster Risk Management (DRM) are overlapping categories of adaptive mechanisms for addressing shocks. A holistic approach for these functionally classified groups of activities (whether state-directed or other) tends to be the most effective approach of governance of systems. The common elements of governance in DRR and DRM include perceptions and information gathering on risks, processing information into knowledge—which then transforms into deliverable projects and activities, efficient planning and management, effective delivery of results. In all these the roles of institutions and their governance is important.

The urgency and severity of the problems of adverse impacts of CC and the roles of actions catered to extreme events as well as the role of CCA has been best illustrated in the IPCC (2012) SREX Report. Some of the highlights are given in Box 1. It is important to quote a major conclusion of the Report (at p. 6 Summary):

...a changing climate leads to changes in the frequency, intensity, spatial extent, duration, and timing of extreme weather and climate events.

The IPCC (2012) SREX Report offers (in about 600 pages) considerable details from various perspectives. These details enable better assessments of risk profiles and impacts of short-term and long-term nature in the absence of adequate preparedness. For example, many tropical regions will have the most increase in number of hot extremes, despite having relatively modest overall or average temperature increases (Anderson 2011; Orlowsky and Seneviratne 2011). These areas also are likely to undergo various other weather extremities, and need urgent attention in CCA and DRR.

CCA can be embedded into DRR but the converse is not always valid. The similarities and differences between the two streams are largely clear, especially after the IPCC (2012) SREX Report; see some of the highlights in Box 1. Most of the differences between DRR and CCA arise, scientifically, from choice of differential spatial and time scales, the knowledge base for further changes, and adoption of norms for each category. Differences between the two categories are better addressed for potential integration when we examine the distinct phases of their applications: pre-disaster preparedness, post-disaster response and mitigation activities, and recovery and reconstruction activities. Adoption of the broader coverage of CCG for some of the common activities such as capacity building is expected to be effective both in terms of cost-effectiveness, and attainment of objectives of both streams of activities.

Box 1: IPCC SREX (2012) Report: Some Highlights

It is with almost certainty that it can be predicted warming temperatures with extremes will occur, and extreme costal high water will also occur. However, the increases in the intensity of droughts and floods are not certain, but somewhat likely.

With or without CC, disaster risk will continue to increase in many countries as more people and assets are exposed to weather extremes. CC changes the magnitude and frequency of some extreme weather and climate events ('climate extremes') for worse.

High levels of vulnerability, combined with more severe and frequent weather and climate extremes, may result in some places.

A balance needs to be struck between measures to reduce risk, transfer risk (e.g. through insurance) and effectively prepare for and manage disaster impact in a changing climate; this balance will require a stronger emphasis on risk diversification and risk reduction. Countries' capacity to meet the challenges of observed and projected trends in disaster risk is determined by the effectiveness of their risk management system. In cases where vulnerability and exposure are high, capacity is low, and weather and climate extremes are changing, transformational changes may be required to avoid the worst disaster losses.

There is substantial overlap of CCA and DRR activities; the differences are largely in terms of scale of activities and their time horizons of operation.

1.5 Equity and Climate Justice

Equity and climate justice issues warrant that victims be compensated reasonably and fairly, and at the same time repetitive claims may be difficult to sustain given the limits of funding from the viewpoint of the compensators, despite potential recurrence of adverse effects of climate change. The message here is the compensation should, after offering priority relief wherever needed, ensure a good degree of adaptation and resilience building. This capacity building activity is a joint responsibility of the victims and others. A judicious mix of resources, instruments, technologies under a joint framework of interventions will be prudent. Green Climate Fund (GCF) in this context augurs well to tailor to the CCA and DRR issues; details will be explored later in this Monograph.

Several studies, including some carried out by the UNFCCC, clarify the fact that CC induces non-uniform adverse impacts on less developed and vulnerable countries. This is even more bothersome for the reasons: these countries have the least contributions in terms of emissions of GHGs contributing to CC (whether these contributions assessed from production systems or consumption systems), and also these have the least capacities to remain resilient in the wake of these unwelcome disturbances. *Accordingly, policies for CCG need to view the implications for an unsustainable living conditions in the least developed and vulnerable countries in the interests of equity and climate justice.*

Analogous logic applies also to vulnerable sections of society in any given region or country, because the consumption patterns of the poor and other vulnerable communities are such these contribute least toward GHGs based on their lower consumption levels of resources, and also these have the least capacities to handle the crises. Thus it is important to recognize these vulnerabilities as well when public policies on social protection mechanisms and those affecting CCG are devised.

Magnitudes of loss and damage, and provision of compensation mechanisms at operationally relevant levels, depend both on the external factors such as CC and on internal factors such as quality of institutions and governance. A cost- minimizing and resiliency-building approach warrants simultaneously consideration of delivery of integrated packages at international, national and sub-national levels in a multi-period setting. Relevant insights and approaches for this purpose are suggested in the subsequent chapters.

References

Anderson, B. T. (2011). Intensification of seasonal extremes given a 2 degree global warming target. *Climatic Change.* doi:10:1007/s10584-011-0213-7.

IPCC. (2007). In M.L. Parry, O.F. Canziani, J.P. Palutikof, P.J. van der Linden & C.E. Hanson (Eds.), *Contribution of working group II to the fourth assessment report of the intergovernmental panel on climate change.* New York: Cambridge University Press. http://www.ipcc.ch/pdf/assessment-report/ar4/wg2/ar4-wg2-app.pdf.

IPCC. (2012). *Managing the Risks of Extreme Events and Disasters to Advance Climate Change Adaptation.* http://ipcc-g2.gov/SREX/images/uploads/SREXSPMbrochure_FINAL.pdf

Kates, R. W., Travis W. R., & Wilbank, T. J. (2012). Transformational adaptation when incremental adaptations to climate change are insufficient. In *Proceedings of the National Academy of Sciences* (USA) (Vol. 109, pp. 7156–7161). www.pnas.org/cgi/doi/10.1073/pnas.1115521109

Orlowsky, B., & Seneviratne, S. I. (2011). Global changes in extreme events: regional and seasonal dimension. *Climatic Change.* doi:10:1007/s10584-011-0122-9.

Rao, P. K. (2011). *International trade policies and climate change governance.* Berlin: Springer.

Rao, P. K. (2012, March 22–23). *Effective governance of climate change adaptation.* Paper presented at the International Conference on Governance of Adaptation, Amsterdam: Vrije University. www.adaptgov.com

Rao, P. K. (2013). *Government Austerity and Socioeconomic Sustainability.* Berlin: Springer Verlag.

Chapter 2
Approaches for Assessing Loss and Damage

2.1 Losses and Damages: Avoidable, Residual, Irreducible and Irreversible

Avoidable loss and damage (L and D) through mitigation and adaptation actions is normally related to both short-term aspects of reducing adverse effects, especially disasters. Mitigation reduces the concentration of GHGs and thus contributes to the reduced potential for climate change and takes a few decades to realize the effects. CCA activities, depending on the scope and scale of relevant activities, cater to reduction of potential impacts of climate change.

Residual L and D is usually the portion that accrues after adjusting for the effects of CCA in the context of adverse impacts. Unavoidable or irreducible damage or inevitable damage is the quantum of L and D, after allowing for the positive effects of CCA and various hazard mitigation (including capacity building, prevention, and governance activities). Since existing mitigation commitments and actions will not prevent dangerous climate change related impacts, residual L and D will follow even after more widespread CCA., The climate change impacts that we are unable to prevent through mitigation and adaptation efforts, will likely be the defining part of the future response to climate change. Warner et al. (2012) illustrate the relevance of the concepts of social vulnerability and social resilience to under-standing how climate change impacts translate into loss and damage for society. Both adaptation deficit and limits to adaptation can result in residual L and D.

Assessment of L and D has to incorporate financial and economic aspects of various hazards and disasters. This Monograph provides a few perspectives in the economic context, and has little of social dimension directly addressed. The key aspect of the social dimension include, but not limited to following, with dire implications for expanding chronic poverty and extreme poverty as a result of disasters: women and children more adversely affected than other sections in any given area, poor are more vulnerable than others in both the vulnerable countries and others (and are more susceptible to fall into chronic poverty and extreme poverty with the implication of least capacity to rise above poverty line), and

K. R. Pinninti, *Climate Change Loss and Damage*, SpringerBriefs in Climate Studies,
DOI: 10.1007/978-3-642-39564-2_2, © The Author(s) 2014

geographically disadvantaged or migratory populations at low incomes suffering more disproportionately. When there is loss of life and/or the likelihood of regaining the original asset and income base (even the meager levels that formed the portfolios of assets and incomes) is very low, the accrual of L and D must be deemed as irreversible. Societies and international arrangements need to pay particular attention to these dire implications (see also Rao 2013).

It is relevant to clarify that not all disasters that occur are contributed by climate change or excessive concentrations of greenhouse gases, and that the international aid processes and resource flows in various channels (bilateral, multi-lateral, aid and relief or other) tend to offset some degree of adverse effects. Thus, all the adverse impacts of climate change may not attract attention for full compensation by the main contributors of climate change. There is a possibility of considerable overlap between the activities of adaptation to climate change and adaptation to natural hazards and disasters. Further development of compensation mechanisms separately and jointly is relevant in this context. As a beginning in the process of devising compensation mechanisms for L and D, it is useful to focus directly on the disaster-proneness and ensure that vulnerable countries are supported for regaining economic, social and environmental resiliency.

2.2 Brief Background at the International Level

In 2007 the Bali Action Plan included concerns by seeking (UNFCCC/CP/2007/6/Add.1) (UNFCCC 2007) 'enhanced action on adaptation', including consideration of risk management and risk reduction strategies, risk sharing and risk transfer mechanisms such as insurance, disaster risk reduction and resources to address loss and damage associated with the adverse impacts of climate change in vulnerable developing countries.

'Loss and Damage' has been introduced under the UNFCCC agenda in 2010 (Decision 1/CP.16). The 2010 Cancun COP launched the work program 'to consider ....approaches to address loss and damage' associated with climate change impacts in developing countries that are particularly vulnerable to the adverse effects of climate change (Decision 1/CP.16 at para 26). The Cancun Adaptation Framework (UNFCCC Decision 1/CP.16) noted that approaches to address loss and damage should consider adverse climatic impacts, including "sea level rise, increasing temperatures, ocean acidification, glacial retreat and related impacts, salinization, land forest degradation, loss of biodiversity, and desertification."

The 2011 Durban COP required (Decision 7/CP.17) "the need to explore a range of possible approaches and potential mechanisms, including an international mechanism, to address loss and damage", with possible recommendations to be considered at the 2012 COP session at Doha. Box 1 provides a summary of the Doha resolutions.

Box 2: COP 18 Decisions on Loss and Damage

Country Parties agree to the following actions, among others:

1. Assess the risk of loss and damage associated with the adverse effects of climate change, including slow onset impacts; identifying options and designing and implementing country-driven risk management strategies and approaches, including risk reduction, and risk transfer and risk-sharing mechanisms; implement comprehensive climate risk management approaches; involving vulnerable communities and populations, and civil society, the private sector and other relevant stakeholders, in the assessment of and response to loss and damage;
2. Identify and develop appropriate approaches to address loss and damage associated with the adverse effects of climate change, including to address slow onset events and extreme weather events, including through risk reduction, risk sharing and risk transfer tools, and approaches to rehabilitate from loss and damage associated with the adverse effects of climate change;
3. Enhance coordination, synergies and linkages among various organizations, institutions and frameworks, to enable the development and support of approaches to address loss and damage, including slow onset events and comprehensive climate risk management strategies, including risk transfer tools;
4. Enhance capacity-building at the national and regional levels to address loss and damage associated with the adverse effects of climate change;
5. Developed country Parties to provide developing country Parties with finance, technology and capacity-building;
6. Establish institutional arrangements, such as an international mechanism, including functions and modalities, ...to address loss and damage associated with the impacts of climate change in developing countries that are particularly vulnerable to the adverse effects of climate change.

Source: www.unfccc.int, Document UNFCCC/CP/2012/L.4/Rev.1.

2.3   Trends in 'Normalized' Disaster Losses

Economic losses from disasters will continue to increase around the world. Since 1981, economic loss from disasters is growing faster than GDP per capita in the OECD countries. This means that the risk of losing wealth in weather-related disasters is now exceeding the rate at which the wealth itself is being created (UNISDR 2011). Economic exposures to disasters are faster than per capita GDP (Mitchell et al. 2012), and adverse impacts of climate change will continue to accelerate incidence of extreme events and their corresponding losses. Low income countries account for large segment of human losses, and middle income countries account for largest economic impacts measured as percentages of GDP.

The magnitude of losses in developing countries is very significant: average annual losses due to disasters were about 9 % of GDP during 1997–2001 (Mahul and Gurenko 2007). Excessive risk impacts and their trends in recent years around the globe suggest the need for better preparedness and loss reduction in all possible scenarios. It is important to note that all losses contributed by disasters can hardly be quantified. A Special Report prepared for G20 Summit 2012 stated (World Bank et al. 2012, p. 13):

1. "Because they are harder to quantify, the indirect consequences are rarely considered, but can have important negative impacts on development achievements and poverty reduction efforts".
2. "The financial consequences of disasters are one of many types of fiscal risk that are faced by governments; because they are difficult to measure, they are often ignored".

Thus, damage estimates of disasters remain underestimated, and the gravity of losses is such that disaster risks need to be integrated into financial resilience assessments of countries, regions, and local areas. Just as the total cost assessment is fraught with data limitations, the quantification of costs of socio–economic vulnerability when extreme events occur remains an area founded on approximations. However, ranking of policies in relation to objectives such as poverty reduction and disaster risk reduction is possible, analytically.

"The negative fiscal impacts of disasters can hamper longer-term growth and economic development.... In times of constrained public budgets, planning ahead for the financial coverage of future disaster costs becomes....necessary ...worldwide" (World Bank et al. 2012).

The above recommendation has both macroeconomic and microeconomic dimensions, in linking various sections of society, their vulnerabilities and aggregative effectiveness of public budgets on socio-economic resilience in addition to fiscal resilience in response to disasters.

Disaster Risk (DR) is a non-linear function, determined by a combined effect of exposure (E), vulnerability (V) and hazard (H):

$$DR = f (E,V,H)$$

The arguments in function f are themselves functions of several parameters. Exposure E is determined by the infrastructure and its access to the poor (such as housing and locations that are less prone to hazards). Vulnerability V can be classified in terms of financial, economic, environmental and social categories of ingredients. V is determined by income levels and socio–economic inequalities, social capital, safety net or other insurance mechanism to protect against severe changes in the economy-environment-hazard frameworks, and sustainable living conditions. Hazard H is an exogenous intervention. The intensity of the effects of

H is influenced by the features that prevail in E and V. Disaster results occur when E and V are both high, and when H itself is high. DR can be minimized largely by addressing V. Vulnerability reduction is a core common element of both CCA and DRR. Vulnerability is an outcome of skewed development processes and scarcity of livelihood options for the poor (IPCC 2012, p. 10). Economic vulnerability refers to the inability of affected individuals and entities to absorb or cushion damages. In this context the role of poverty as a precursor to vulnerability to external shocks—physical and economic- is rather well-known. Life risks and mortality with demographic inequities are accentuated in poorer areas and segments of populations. Also, disasters lead to livelihood risks and to chronic poverty, besides quality of life risks and accentuated inequalities.

As stated in the G20 Special Report (World Bank et al. 2012): (*a*) *disasters are the result of the interaction between hazards, assets, and vulnerability; and,* (*b*) *integration of disaster risk information in national fiscal risk management frameworks can help improve the fiscal resilience of countries.* In general, the dichotomy between fiscal stability/sustainability (with or without spending cuts) and the corresponding objectives with the advent of extreme events needs to be visualized. A balanced approach of interventions for addressing both the *ex ante* and *ex post* scenarios (relative to an extreme event) is called for. Reduction of socioeconomic vulnerability merits priority attention in both scenarios to ensure system stability.

There are discernible differences in the ongoing and potential impacts of climate change around different regions of the world. Much of Sub-Saharan Africa is forecast to experience a reduction in the length of the growing season, with a reduction of at least 20 % by 2100. East Africa is expected to suffer a 20 % loss in maize yields, while West Africa is predicted to suffer a staggering 90 % cut in its bean production. This reflects the likelihood that in a 4 °C world more than one-third of current cropland in east and southern Africa would be unsuitable for cultivation (Thornton et al. 2011).

Losses follow geophysical changes in most cases (Bouwer 2011), and the roles of changes in people-centric vulnerability and exposure. Thus, it is often interpreted (see also Neumayer and Barthel 2011) that disaster losses are not necessarily increasing (as a percentage of GDP in many countries) around the world if we adjust for these contributory factors. An implication of such assessments could sidetrack the ever increasing adverse changes in the climate system and its effects, besides several people-centered activities that contribute the changes in the climate system. Even in those countries that have seen loss and damage (as a percent of respective GDPs) will have less scope enhancing or sustaining their standards of living if no pro-active measures are initiated to address the disaster-related infrastructural imperatives in the next few years. The key issues are how to address the adverse effects at national and international levels, introduce an element of climate justice, and effective governance. It is a joint and simultaneous responsibility for both the groups of countries: the vulnerable and the others. If the potential victim regions or countries do less than the optimum efforts to gear up to the tasks, any amount of resource transfer or compensation may not solve the real problems on a sustainable basis.

There is a range of challenges which should be taken into account UNFCCC (2012 Document FCCC/TP/2012/1):

(a) Capturing the scope and extent of direct and indirect losses as well as the growing interconnectedness of impacts (such as cascading effects);
(b) Further clarification of the strengths, weaknesses and limitations of the available methods and tools with a view to avoiding misunderstandings and misuse –particularly in the context of uncertainty (climatic and non-climatic); and,
(c) Enhancing methods and tools for assessing the risks from slow onset changes, such as sea level rise, salinization or the degradation of ecosystems and ecosystem services.

L and D assessments associated with climate change need to include both slow onset adversities such as sea level rise, and sudden extreme events such as floods. However, there is no clear definition of loss and damage under the UNFCCC or any other relevant international agreement thus far. UNFCCC (2012a) reviews in its Technical Paper a substantial part of analytical methodologies that could be potentially relevant for assessing losses and damages under varying risk factors in relation climatic and non-climatic factors and their dynamics over time. After a brief review of 18 approaches the Technical Paper narrowed down six of these for further attention:

Catastrophe risk models CATSIM
Comprehensive Approach for Probabilistic Risk Assessment CAPRA
Integrated Assessment Models IAMs
Scenario-driven approaches
UK Climate Change Risk Assessment CCRA
World Risk Index WorldRiskIndex

It is not proposed to dwell upon the details in this Monograph. Perhaps it helps to note that with longer time horizons that are applicable to slow onset events, low or zero discount rate becomes relevant for evaluating future costs and benefits or related portfolios of interventions/investments. Technically, Laplacian Operators (infinite time horizon models) become applicable for computational analysis, but not seen yet in applications. None of these constitute ready-to-use toolkits, however. Besides, financial and economic dimensions of analysis constitute merely one important component of total assessment.

In one of the studies- relevant for an understanding of time horizons of CCA activities, Hallegatte (2009) identified major sectors, time-scales and applicability of no-regrets criteria:

Category A: High Exposure to climate risks
Water infrastructure (dams, reservoirs and such other hardware) (30–200 years)
Land use planning (especially coastal and flood prone) (about 100 years or more)
Coastal flood defenses (about 50 or more years)

Category B: Medium Exposure
Buildings and Housing (30–150 years)
Category C: Low Exposure
Transportation (30–200 years)
Urban structures (about 100 or more years)
Energy production (20–70 years)

The recommended priority sectors and sub-sectors for no-regret strategies include the following:

Development of climate resilient crops
Early warning and evacuation systems
Improvements in public health systems
Institutionalization of perspective planning
Enhancement of water use efficiency
Sustainable land-use planning

These measures, in combination with the critical elements of comprehensive infrastructure development for disaster reduction and management, constitute some of the key elements of foundations and perspectives toward reduction of L and D.

UNFCCC (2012b) report provides a literature review on a broad range of approaches to address loss and damage associated with the adverse effects of climate change. Among these are risk reduction, risk retention, and risk transfer; issues of slow onset phenomena and enabling environments to reduce loss and damage have also been briefly examined, as well as regional priorities around the world. The Report provides an array of approximate qualitative listing of categories costs and benefits potential interventions to address the impending and possibly accelerating adverse impacts. These tabulations are of some use but no numerical estimation or methodological basis can be founded yet. Besides, most analyses, whether arising from the UNFCCC or others, depict deficiencies in not being attentive to the critical role of endogenous responses of various stakeholders to exogenous interventions and to the emerging information about climate change and its applicable effects (as perceived by these actors).

Compensation for L and D remains an issue in the areas of equity and climate justice across communities and nations. Least developed countries (LDCs) are also the least contributors to climate change but ironically the most adversely affected by the phenomenon contributed by other countries. This is a global scale externality and cannot go on uncompensated. It has been suggested UNFCCC (2012d) suggested (at para 72) it is important to find ways of working through the CCA and CCM to facilitate the 'development of a clear narrative for concept of loss and damage, "which is important for prioritizing addressing the issue, including for identifying the means of addressing loss and damage."

Financial and economic elements for assessing L and D.

A 'working definition' (www.loss-and-damage.net) suggests that damage is the set of negative impacts that can be remedied or restored, and loss is the set of negative impacts that cannot be restored to ex-ante scenarios.

There is also the case that a continuum exists between the two, and also between 'slow onset' events and 'sudden extreme' events. A mix of elements applies in each location and time interval, some of which is predictable but much is not entirely predictable at that scale and time specificity.

Other economic elements of assessing L and D include:

Valuing life
Loss of economic potential and productivity
Loss of jobs and business income

At country or regional level as well local levels the implications of incidence of disasters need to be assessed in terms of the potential accrual of chronic poverty and extreme poverty. Although there are large uncertainties in information, cause and effect relationships and stakeholder responses over time in terms of pro-active or reactive measures and capacity building activities and other aspects of adaptation. The role of innovation in technologies, infrastructure planning and development, governance mechanisms remain critical; potential L and D consequences of hazards and disasters (both extreme events and slow onset phenomena) are related to these features.

2.4 Loss and Damage Features and Estimates

People around the world have to face the reality of climate variability and its adverse consequences; most regions tend to experience increased variability. This is not the same thing as mere warming of mean surface temperatures and these variations induce effects for which almost no system is fully prepared for. Direct economic losses relative to national income in developing countries were about twice the corresponding proportions of developed countries during the past quarter century; "planning for both current climate variability and longer-term shifts in climate patterns can help smooth pathways and cushion the negative impacts of loss and damage in the future" (Warner and Zakieldeen 2012).

The following are some of the major sources of assessments of loss and damage, merely to to illustrate trends over time and spatial priorities and magnitudes involved.

1. The nodal agency CRED analysis of indicates (see Guha-Sapir et al. 2012) that the number of victims of disasters has been on the rise over the years: the global total in 2011 stood at 244.65 million people (of which 211.16 million are from Asia); the corresponding average for the period 2001–2010 was 231.95 (and 207.16 million for Asia). The estimates of damages (all at 2011 US$ in billions, not assessing the value of life for lost lives) for the global total in 2011 stood at US$366.12 (with major segment contributed by Asia at US$276.03, whereas the corresponding averages for the period 2001–2010 stood at US$109.35 (with Asia at US$41.61).

2. According to the Harmeling and Eckstein (2012), the *Global Climate Risk Index* ranks Honduras, Myanmar, Nicaragua, Bangladesh, Haiti, and Vietnam as the countries most affected by extreme weather events from 1992 to 2011. In most affected countries (1992–2011) were developing countries in the low-income or lower-middle income country group. In total, more than 530,000 people died as a direct consequence from almost 15,000 extreme weather events, and losses of more than 2.5 trillion USD (in purchasing power parity PPP) occurred from 1992 to 2011 (USD 1.68 trillion overall losses in original values). More than 530,000 people died as a direct consequence of almost 15,000 extreme weather events, and losses of more than USD 2.5 trillion (in PPP) occurred from 1992 to 2011 globally.

3. Climate change and pollution related to carbon-dioxide emissions are reducing the world's gross domestic product (GDP) by 1.6 % a year, about \$1.2 trillion, according to DARA International (2012) Report *Climate Variability Monitor*. Climate change may cut GDP in some developing nations by as much as 11 % by 2030, and the worldwide net losses of GDP could be about 3.2 % due to the effects of carbon emissions and climate change. Those losses far exceed the cost of reducing emissions, which the report estimated at about 0.5 % of GDP over the next decade. Climate change was responsible for about 5 million deaths in 2010, including 400,000 related to hunger and diseases and 4.5 million from air pollution, according to the report. Data suggest the increasing frequency and intensity of extreme weather events exacerbating the socio–economic problems of the Least Developed Countries (LDCs) and Small Island Developing States (SIDS) and contributing to political instabilities and chaotic migration.

4. CRED reports and data highlight a few events. The disaster that made the most victims in 2011 was the flood that affected China in June, causing 67.9 million victims. Furthermore, China was affected by a drought from January to May (35 million victims), a storm in April (22 million victims) and another flood in September (20 million victims), further contributing to a total of 159.3 million victims in China in 2011, a figure representing 65.1 % of global reported disaster victims. Droughts and consecutive famines made many victims in Ethiopia (4.8 million), Kenya (4.3 million) and Somalia (4 million). When considering the population size of the country, 42.9 % of Somalia's population was made victim of natural disasters in 2011, mostly due to drought.

In 2011 the number of disaster victims has increased significantly relative to the average of the previous decade (Guha-Sapir et al. 2012). This increase is explained by the larger impact from hydrological disasters. Hydrological disasters caused 139.8 million victims in 2011—or 57.1 % of total disaster victims in 2011—compared to an annual average of 106.7 million hydrological disaster victims from 2001 to 2010. In 2011, 66.8 % of global hydrological disaster victims were from floods and wet mass movements in China.

2.5 Drought and Slow Onset Adverse Effects

2.5.1 Slow-Onset Events

There are differences in the manner in which the disaster risk reduction community and the climate change community conceptualize the term slow onset (Siegel 2012):. The disaster risk reduction community views slow onset hazards as disasters that unfold slowly over months or several years. In the climate change process slow onset time scales are counted in years and decades (even longer time horizons).

The impacts of slow onset events are already being felt in all regions and are exacerbating extreme weather events, but that there is limited readiness to address these impacts, in terms of the institutions and capacities in place at all governance levels. The existing gaps related to knowledge on and tools for addressing such impacts, in comparison with current knowledge on and available tools for addressing extreme weather events, were highlighted at all of the expert meetings coordinated by the UNFCCC; for details, see UNFCCC (2012c). Managing the risks associated with climate change, in particular the risks associated with slow onset events, requires long-term planning and institutional arrangements with appropriate legislation and policies, as well as reliable governance structures across sectors and levels, supported by timely, quality information and sustainable commitments to providing financial resources.

There is an urgent need to improve the understanding of the characteristics of slow onset events, including the linkages with extreme weather events, definitions of baselines for slow onset events, potential tipping points, the capacity and skills needed for quantifying losses, and what types of approach are necessary. Such an improved understanding would lead to raised awareness of the magnitude of the loss and damage resulting from incremental climatic processes, especially among policymakers, and facilitate a clarification of the necessary enabling environment, such as regulatory frameworks, policies and institutional structures. This would, in turn, facilitate the avoidance of institutional fragmentation in addressing slow onset events. While some successful practices were introduced, 21 discussions at the expert meetings for the Latin American region and SIDS highlighted the limitations of using infrastructural measures to address slow onset events at the appropriate temporal and spatial scales. Discussions on ocean acidification and loss of biodiversity as a result of slow onset events drew attention to the permanence of the loss of biodiversity and its impact on livelihoods for current and future generations, which conventional adaptation measures, often project-based approaches, have limited effectiveness in tackling. The relatively short-term cycle of donor funding poses challenges in this regard in terms of enabling the long-term nature of the action often required to address slow onset events.

2.5.2 Drought

The impacts of drought are increasing in magnitude and complexity due to the effects of a changing climate. Unlike other natural hazards such as storms, earthquakes and floods, which occur with a specific period of time and result in concrete damages, drought emerges slowly and quietly and lacks highly visible and structural impacts. When does it begin, when does it end? Geographically speaking, where are the limits of its spatial impacts?

The lack of standardization in drought hazard characterization contributes to the problem of attributing definitive losses. Even if drought information has improved and the methodology applied in EM-DAT has been strengthened, data still remain inconsistent because of the complexity of droughts, especially in terms of measuring the direct human impact. Indeed, the impacts of drought may endure for years, and providing a strict spatial definition is difficult due to the spatial patterns of droughts and the localized nature of precipitation.

Understanding the complex impacts of drought could be the key to enhancing drought mitigation and preparedness. "Data on disaster losses in Africa is low", highlights the UNISDR in its Briefing Note no. 4, entitled "Effective measures to build resilience in Africa to adapt to climate change". This fact does not lessen the evidence showing that GDP growth in African countries is under threat from the impact of natural hazards, particularly agricultural drought. "Drought is predictable and does not happen overnight. Therefore, it should not claim lives nor lead to famine, which results when drought is coupled with policy failure or governance breakdown or both."

The Global Assessment Report on Disaster Risk Reduction (GAR11) of the UNISDR highlights improvements in early warning, preparedness and response. "The massive mortality from Sub-Saharan African droughts in the 1970s has not been repeated". However, compared to other hazards, risks associated with drought remain poorly understood and badly managed, particularly in some African countries. To avoid these gaps, UNISDR released "Drought contingency plans and planning in the Greater Horn of Africa" in early 2012. According to the IPCC, the Sahel and West Africa are among the most vulnerable regions to future climate fluctuation. The food crisis is becoming chronic, because the majority of the population depends on agriculture for the livelihoods. "Maybe more than any other disaster risk, drought risk is constructed by economic decisions and social choices".

Although there has a great deal of understanding on the potential for integrating CCA, DRR, SD and poverty reduction, this is not yet translated into practice. For example, an evaluation study of the UNDP (2011) concluded:

> Although the UNDP strategic priorities acknowledge the links between poverty reduction, SD and DRR, these strategies are not systematically implemented, and the DRR strategy should be revised to more directly address CCA.

The Hyogo Framework of Action (HFA) recognizes the role of institutions and states in the first pillar of action: "Ensure that disaster risk reduction is national and a local priority with a strong institutional basis for implementation". However, the multi-dimensional nature of institutions needs to be reflected in the ongoing updating approaches for the post-2015 development agenda frameworks.

> Integrating CCA into the HFA is a relevant ingredient of post-2015 framework. The UNFCCC process needs to enable such integration, since CCA currently is largely in its mandate. Policy documents of the UNFCCC dating back some years advocate the role of integration but the premise has not been brought into practice so far in a significant manner (Rao 2013a).

Cost-effective forward-looking mechanisms for reducing L and D include the following, besides effective disaster governance: Roles of climate services, Role of extensive use of information and communication technologies (ICTs), Crop insurance, Property and casualty insurance, and Life insurance. None of these policy mechanisms are currently affordable for most of the vulnerable countries (especially the Least Developed countries, LDCs), and the new international mechanisms (whether to be called market mechanisms or other international institutional arrangements) will be most useful if they enable affordability of these technologies and other resources with considerable support from the developed countries. This will create relevant infrastructure, combined with a lot more dynamic international trade policy –especially in environmental goods and services.

There is a significant scope to examine effective (including cost-effective) options about to address L and D in terms of Risk Reduction (Prevention and Governance), Risk Retention and Risk Transfer. This Monograph suggests that the international mechanisms being devised for compensation for L and D must assess the cost-sharing implications under each of these options so that the overall effect of resource support enables the vulnerable less developed regions to form and upgrade relevant national and sub-national systems to cater to the L and D prevention, mitigation and governance.

It is useful to recognize the role of the implicit and explicit relationships governed by the fundamental relationship:

$$\text{Loss and Damage} = f(\text{Hazard}, \text{Vulnerability}, \text{Adaptive Capacity})$$

Expanding CCG agendas suggest the potential for embedding DRR into CCA frameworks into those of sustainable development as well as the emerging frameworks of post-2015 development agendas that are in progress.

Identifying country priorities (UNFCCC 2012d) include an assessment of L and D starts in identifying the assets that are at risk due to the adverse effects of climate change. "The multifactor nature of this issue poses a challenge in building bridges between stakeholders from different disciplines when trying to integrate efforts. There is a need to integrate people working on, inter alia, adaptation, disaster, crisis and environmental management, as well as development, not only for technical cohesion but also for building the environment in which their efforts can be enhanced in a coherent manner."

Detailed micro-studies including household surveys in five countries Bangladesh, Bhutan, The Gambia, Kenya, and Micronesia indicate that adverse impacts of both slow-onset and extreme events impact households currently and considerable effects are felt in relation to their vulnerability (Warner et al. 2012). Losses and damages occur because of the following factors:

Insufficient and ineffective adaptation measures
Cost-effectiveness or benefit-cost ratios not attractive
The measures help in the short-term but cause adverse long-term consequences
Paucity of resources that limit the design and implementation.

The study suggested the need to support community level assessments of risks and their building resilience with enhanced support for local sustainable development and reduction of socio–economic vulnerabilities. The screening of CCA (and some in joint frameworks with CCM) projects and activities must be subject to the L and D criterion, and advance pro-poor gender-sensitive agenda. This reduces the incidence of L and D both in the short-run and in the long-run. Thus, mainstreaming disaster reduction in development and implementation of projects and activities remains a high priority.

Doha outcome that seeks to address some of the concerns of the least developed countries and small island countries is to establish in about one year "institutional arrangements, such as an international mechanism" to address L and D due to the adverse effects of climate change in particularly vulnerable developing countries. Some of the relevant perspectives in this regard are offered in the chapters to follow.

References

Bouwer, L. M. (2011). Have disaster losses increased due to anthropologic climate change? *Bulletin of the American Meteorological Society, 92*, 39–46.

DARA International. (2012). Climate variability monitor, Madrid: DARA International www.daraint.org

Guha-Sapir, D., Vos, F., Below, R., & Ponserre, S. (2012). *Annual disaster statistical review 2011: The numbers and trends.* Brussels: CRED. Available at http://www.cred.be/sites/default/files/ADSR_2011.pdf

Hallegatte, S. (2009). Strategies to adapt to an uncertain climate change. *Global Environmental Change, 19*, 240–247.

Harmeling, S., & Eckstein, D. (2012). Global climate risk index 2013: Who suffers most from extreme weather events? Weather-related Events in 2011 and 1992 to 2011, Briefing paper, Bonn: Germanwatch.

IPCC. (2012). *Managing the Risks of Extreme Events and Disasters to Advance Climate Change Adaptation;* available at http://ipcc-g2.gov/SREX/images/uploads/SREXSPMbrochure_FINAL.pdf

Mahul, O., & Gurenko, E. (2007). The macro financing of natural hazards in developing countries. In World bank policy research working paper, p. 4075. Washington, DC: The World Bank.

Mitchell, T., Mechler, R., & Harris, K. (2012). *Tackling exposure: placing disaster risk management at the heart of national economic and fiscal policy.* London: CDKN.

Neumayer, E., & Barthel, F. (2011). Normalizing economic loss from natural disasters: A global analysis. *Global Environmental Change, 21,* 13–24.

Thornton, P. K. et al. (2011). *Agriculture and food systems in sub-Saharan Africa in a 4 °C + world* (pp. 120–123). The Royal Society. http://www.agriskmanagementforum.org/farmd/ sites/agriskmanagementforum.org/files/Documents/ag%20and%20food%20systems%20in% 20SSA%20in%20a%204C+%20world.pdf

Rao, P. K. (2013). *Governance of disaster reduction,* Working Paper, Geneva: UNISDR.

UNDP. (2011). *Guidelines and lessons for establishing and institutionalizing disaster loss database.* New York: UNDP.

UNISDR. (2011). *Global assessment report on disaster risk reduction (GAR2011).* Geneva: UNISDR.

UNFCCC. (2007). *Report of the Conference of the Parties on its 13th Session,* Bali, December, 2007, Report FCCC/CP/2007/6/Add.1, Bonn: UNFCCC.

UNFCCC. (2012a). *Current knowledge on relevant methodologies and data requirements as well as lessons learned and gaps identified at different levels, in assessing the risk of loss and damage associated with the adverse effects of climate change,* Bonn: UNFCCC Technical Report FCCC/TP/2012/1.

UNFCCC. (2012b). *A literature review on the topics in the context of thematic area 2 of the work programme on loss and damage: a range of approaches to address loss and damage associated with the adverse effects of climate change,* Bonn: UNFCCC Paper FCCC/SBI/ 2012/INF.14.

UNFCCC. (2012c). *Report on the regional expert meetings on a range of approaches to address loss and damage associated with the adverse effects of climate change, including impacts related to extreme weather events and slow onset events,* Bonn: UNFCCC Report FCCC/SBI/ 2012/29

UNFCCC. (2012d). *Report on the expert meeting on assessing the risk of loss and damage associated with the adverse effects of climate change,* Bonn: UNFCCC Document FCCC/SBI/ 2012/INF.3.

Warner, K., Van der Gees, K., Hreft, S., & Haq, S., Harmeling, S., Kusters, K., & de Sherbinn (2012). *Evidence from the frontlines of climate change: Loss and damage to communities despite coping and adaptation,* Policy Report #9, Bonn: UN University Institute for Environment and Human Security (UNU-IEHS).

Warner, K., & Zakieldeen, S. A. (2012). *Loss and damage due to climate change: An overview of the UNFCCC negotiations.* London: European Capacity Building Initiative.

World Bank., GFDRR., & The Government of Mexico. (2012). *Improving the assessment of disaster risks to strengthen financial resilience. In A special joint G20 publication by the Government of Mexico and the World Bank.* Washington, DC: GFDRR.

Siegele, L. (2012). *Loss & Damage: The theme of slow onset impact.* London: Routledge.

Chapter 3
International Environmental Law

3.1 Introduction

The sources of international law arise from a number of origins and comprise both soft law (indicative but non-binding, as in the Hyogo Framework of Action fr resilience building and disaster reduction), and hard law (binding, as in the Montreal Protocol to the Vienna Convention on the Depletion of Ozone Layer). Lack of identifiability and accountability in international environmental damage remains a serious obstacle in the governance of the global commons (Rao 2002). The choice of policy instruments cannot be independent of the specifications of the relevant institutional configurations associated with the design and implementation of these instruments. Doha resolutions on compensation for loss and damage are just a prelude toward further development of legally binding agreements. However, a workable win–win framework of international cooperation can offer a good deal of relief on a track of sustainable alleviation of loss and damage in many countries. Whereas the backdrop of well-structured legal instruments is an enabler for some of the important interventions in mainstreaming disaster and loss reduction, a purely litigious route to provide and receive support in the vulnerable regions is not necessarily effective in terms and timeliness and effectiveness. This chapter enunciates a few guiding principles, recognizes the prevalence of substantial legal vacuum in related aspects, and suggests an expeditious international agreement in the interests of all.

3.1.1 Legal Vacuum

Global common resources may be broadly categorized into two rather non-overlapping types (Rao 2002):

(a) Common property resources *res communis,* and
(b) Open-access resources *res nullius.*

K. R. Pinninti, *Climate Change Loss and Damage,* SpringerBriefs in Climate Studies,   25
DOI: 10.1007/978-3-642-39564-2_3, © The Author(s) 2014

Property rights (PR) as well as liability rules (LR) apply in (a) and PR do not apply in (b).

Legal vacuum *non liquet* exists in a large segment of the global common resources, and lack of LR remains a major problem when negative externalities result from actions/inactions of parties.

In general, international environmental law tends to be forward looking (Rao 2002) and seeks to address *erga omnes* obligations of the States: obligations owed to multiple States and can be invoked separately or jointly by States. Provision of an *amicus curiae* role for non-party States is also suggested as a relevant mechanism to influence the conduct of States to minimize their negative externalities on others (Rao 2002). Recognizing multiple factors such as socio-economic vulnerability and multi-level governance of disaster reduction remain among the forefronts of shared responsibility within and across nations.

3.2 Legal Foundations

One of the important legal strands is the 1987 US Third Restatement of Foreign Relations Law (section 601) deals with state obligations for transboundary damages: "a state is obliged to take all necessary precautionary measures where an activity is contemplated that poses a substantial risk of a significant transfrontier injury".

Later, the 1991 International Law Commission Draft Articles (at Draft Article 24) states that "if the transboundary harm proves detrimental to the environment of the affected State...the State of origin shall bear the costs of any reasonable operation to restore, as far as possible, the conditions that existed prior to the occurrence of the harm. If it is impossible to restore these conditions in full, agreement may be reached on compensation, monetary or otherwise, by the State of origin for the deterioration suffered".

As part of the soft law, the 1992 Rio Declaration Principle 2 states:

> States have, the sovereign right to exploit their own resources pursuant to their own environmental and developmental policies, and the responsibility to ensure that activities within their jurisdiction or control do not cause damage to the environment of other States or of areas beyond the limits of national jurisdiction.

If general legal responsibility for climate change damage is established, such obligation also covers adaptation measures (and costs) as direct damage prevention measures (Tol and Verheyen 2004). Are there relevant precedents, even the emerging law is not necessarily founded on this jurisprudence? We have a few but illustrations.

In the Barcelona Traction Co. case (*Belgium v. Spain*, 1970 ICJ, 3, 32; February 5, 1970) the International Court of Justice (ICJ) recognized the existence of "obligations of a state towards the international community as a whole", and not necessarily confined to the consequences of actions and inactions applicable to the contending parties only.

National and international legal regimes of relevance for climate change damage liability have been documented in Lord et al. (2012), with main focus on the role of greenhouse gas emissions at various levels and scales. In the international perspectives that combine various types of adaptation and mitigation measures within and outside the UNFCCC framework, with potential alternative scenarios of combinations of policies, the future of international climate change law can be examined in an integrated approach. One of the main directions could emerge in seeking a reasonable mix of coordinated efforts to CCA, CCM and DRR mechanisms as well as their governance.

Among some of the major international agreements that explicitly provide for compensation of damages is the 1999 Basel Protocol on Liability and Compensation for Damage Resulting from the Transboundary Movements of Hazardous Wastes and their Disposal (1999 Basel Protocol). This provides a comprehensive regime for liability and for adequate and prompt compensation for damage resulting from the transboundary movement of hazardous wastes, based on both strict and fault liability. The main features of the 1999 Basel Protocol are similar to those of other liability conventions. It imposes joint and several strict liability (with exemptions), and covers damage relating to loss of life or personal injury; loss or damage to property; loss of income; measures mitigating the damaged environment; and costs of preventive measures. Article 14 of the Basel Convention provides that the parties shall consider the establishment of a revolving fund to assist on an interim basis in case of emergency situations to minimize damage from accidents.

The Protocol only applies to damage due to an incident occurring during a transboundary movement and disposal of waste. Rather than assigning liability to a single operator, there is the potential to hold generators, exporters, importers and disposers liable at different stages of the movement of the transboundary waste. It is useful to note that, as in most cases of liability law, implementation is fraught with excessive litigation and costs. It is least expected that the outcome of Doha mandate leads to such cumbersome routes for provision of climate justice and compensation for loss and damage.

Another (though regional) agreement for attention is the 2003 Protocol on Civil Liability and Compensation for Damage Caused by the Transboundary Effects of Industrial Accidents on Transboundary Waters, adopted by the United Nations Economic Commission for Europe (2003 Kiev Protocol). The involvement of States, industry, the insurance sector and intergovernmental and non-governmental organizations in the negotiating process was rather unusual and some of these features deserve further attention in the context of development of post-Doha agreements for loss and damage compensation.

The United Nations Convention on the Law of the Sea also appears to uphold the requirement of equal access (Article 235, paragraph 2) reads: States shall ensure that recourse is available in accordance with their legal systems for prompt and adequate compensation or other relief in respect of damage caused by pollution of the marine environment by natural or juridical persons under their jurisdiction.

Loss and Damage issue may be seen also from the viewpoint of equity and justice. Governance of climate change and its adverse impacts is a global collective public good, and is a theme founded on challenging and innovative agendas with unprecedented frameworks.

In general, the norms of 'no harm' rule and state responsibility, and obligations before harm occurs, remain guiding principles for the development of the background toward compensatory mechanisms.

Article 4.1 (b) of the UNFCCC charter obliges all Parties to "formulate and implement national or regional programmes containing measures to mitigate climate change and measures to facilitate adequate adaptation to climate change". Thus, adaptation is not a voluntary undertaking but a substantive obligation on all Parties with a view to reducing future climate change damage. However, there is uncertainty as to what constitute "adequate" adaptation measures and when and exactly how the obligation must be met (Tol and Verheyen 2004). We still need a functionally meaningful, beyond conceptually elaborate, working definition of CCA. Despite considerable efforts and activities to adopt CCA in several areas around the world, adaptation deficit remains too significant, mainly because of the scale of resource requirements. This paves the way for disaster-proneness as a contributory factor though not necessarily the main factor.

International law supports mechanisms negotiation and of cooperation. International environmental law, protects global commons such as the oceans and the atmosphere, and is an example for such "*erga omnes*" obligations (an obligation which is owed to a multitude of states and can thus be invoked by these jointly or individually). Legal duty to pay compensation beyond adaptation support (see also Verheyen and Roderick 2008; and Verheyen 2005) may be very relevant, but the two streams can be and should be integrated for greater effectiveness in reducing loss and damage.

As the international focal point for activity on global climate change activities, the UNFCCC process should take a lead role in inspiring, coordinating and synthesizing action to address loss and damage to slow onset climate change hazards.

3.3 Relevant Articles of the UNFCCC

The UNFCCC charter in its Preamble at para 9 refers to the role of 'no harm rule' and state responsibility with implicit obligations before harm occurs. The operative provisions of the Convention begin by defining both the 'adverse effects of climate change' as well as the term 'climate change' itself. Adverse effects of climate change are

> …changes in the physical environment or biota resulting from climate change which have significant deleterious effects on the composition, resilience or productivity of natural and managed ecosystems or on the operation of socio-economic systems or on human health and welfare… (Article 1.1)

UNFCCC Article 4 states:

"Developed countries shall assist developing countries that are particularly vulnerable to
the adverse effects of climate change in meeting costs of adaptation..........."
"Parties are to provide funds, insurance, and the transfer of technology to actions to meet the
specific needs and concerns of developing country Parties, arising from the adverse effects
of climate change, especially countries with areas prone to natural disasters".

The UNFCCC charter augurs well to direct national and international actions to
prevent, mitigate and govern the relevant activities—although the governance
aspects remain extremely weak. Lack of scientific knowledge is no excuse for not
acting, especially where there is the potential for irreversible damage (Article 3.3).
Cooperation is required in preventing and rehabilitating loss and damage to those
areas affected by hazards such as drought and desertification (Article 41(e)), and
developed countries are required to assist particularly vulnerable developing
countries in meeting the cost of adapting to the adverse effects of climate change
(Article 4.4). The nature of the costs to be covered and support provided includes
transfers of technology (Articles 4.3 and 4.5) and developed countries must take
the lead in fulfilling their commitments under the Convention (Article 4.7).

UNFCCC Article 4.8 states:

Parties are to provide funds, insurance, and the transfer of technology to actions to meet the
specific needs and concerns of developing country Parties, arising from the adverse effects
of climate change, especially countries with areas prone to natural disasters.

If only the stated provisions of the UNFCCC are reasonably well acted upon at the
scale and pace of implementation, we could have already seen some good results in
mainstreaming CCG, and DRR. The success of the Montreal Protocol to the Vienna
Convention on Ozone Depleting Substances could have been paralleled but the
situation is far from satisfactory, with worsening global warming and climate change
scenarios, and severe adverse widespread effects. Thus it is time to move into more
transformational yet pragmatic aspects of dealing with CCA, disaster reduction, and
effective provision of measures for loss and damage compensation.

National Adaptation Plans of Action (NAPAs) in Least Developed Countries
(LDCs). Are guided and supported by the UNFCCC preparation of National
Adaptation Plans of Action (NAPAs) in these countries and these plans enable
them to obtain assistance from the UNFCCC, UNDP and the World Bank/GEF.
The resources are too meager to be effective in some of the objectives and goals,
and governance mechanisms lack quality.

The relevant guidelines for the preparation of NAPAs were originally designed in
2001 by the UNFCCC (document FCCC/CP/2001/13/Add.4). These include, *inter
alia*, participatory process, multidisciplinary approach, complementary to other
relevant activities, country-driven approach, cost-effectiveness, sound environmental management, simplicity and procedural flexibility. In practice, some of the
countries adopt a perfunctory approach to participation of stakeholders, and this
aspect is one of the weakest elements of the current guidelines and implementation

of NAPAs. Very few guidelines exist for the governance of the NAPAs, and the scale of operations remains too small relative to the realistic needs of CCA.

The ILC has been developing guidelines and frameworks under its focus and projects dealing with "international liability"; it has set forth articles on "International liability for injurious consequences arising out of acts not prohibited by international law" (see ILC 1998 Report of the Working Group).

International climate agreements are rather conspicuously weak in their compliance and dispute resolution mechanisms. The UNFCCC as a pivotal entity may have to upgrade the system, similar to the WTO Dispute Settlement Understanding.

Dispute Resolution Mechanisms

Article 14 of the UNFCCC charter provides brief rules for dispute settlement. If two or more Parties have a dispute concerning the interpretation or application of the UNFCCC they must seek to settle the dispute through negotiation or any other peaceful means of their own choice (UNFCCC Article 14.1).

If one Party has notified another Party that a dispute exists between them and the Parties have not been able to settle the dispute after 12 months the dispute shall be submitted to conciliation at the request of one of these Parties (UNFCCC Article 14.5).

According to UNFCCC Article 14.6 the conciliation commission is formed by an equal number of members appointed by each Party involved in the dispute. The conciliation commission's award is recommendatory, not legally binding, but Parties are to consider it in good faith (Article 14.6); the award would not be legally binding, it could involve clarifying legal issues and might help advance development of international law on climate change (FIELD 2012a, b). In the UNFCCC and Kyoto Protocol negotiations Parties seem to have preferred to address issues through the negotiating process rather than start a dispute settlement process. Article 14.7 states that the COP shall adopt "additional procedures" relating to conciliation as soon as practicable in an annex on conciliation. The COP has not yet done so.

Under UNFCCC Article 14.1 Parties can choose negotiation or any other peaceful means of their own choice. As noted above, if this does not resolve the dispute within 12 months one of the Parties can trigger conciliation (Article 14.5).

Parties can declare that they accept compulsory dispute settlement through the International Court of Justice (World Court) or arbitration, but very few Parties have done so (see Article 14.2 (a) and (b)). Arbitration usually results in a binding decision. As noted above, the COP has not yet adopted procedures for arbitration.

A claim for compensation for environmental damage affecting *res nullius* could be raised by an umpire-like global trust body, leading to global trusteeship (Rao 2002). This trust doctrine follows the functions those arise similar to the public trust doctrine often used in developed economic systems.

In general, it is preferable both in the interests of effective international law and its practice that the international community comes forward (in minimal time) with a new agreement (in the form of a Protocol to the UNFCCC) to establish mechanisms for CCG and compensation for L and D. This would avail the established norms under the 1968 Vienna Law of Treaties: role and interpretation of lex specialis and lex posterior. The former is the relative overriding nature of provisions under a specialized agency (relative to general agreements) and the next aspect is that agreement devised later in time trump similar agreements made earlier in time—provided the parties to the relative comparisons are the same (see also Rao 2002).

3.4 Laws for Loss and Damage

The estimates of loss and damage due to climate change (whichever source used) are very high; these are too high to be paid by the main contributors of climate change (historically and/or currently). The purpose of deriving more precise estimates is to seek some sense of proportionality of cause-and-effect and loss-and-damage so that vulnerable countries obtain compensation as partial relief. Since full *ex ante* restoration of assets and resources is not contemplated under any of the current and proposed laws at the international levels, it is prudent to seek compensation mechanisms resources that cater to short-term and long-term aspects of DRR and CCA in the vulnerable countries that have low levels of infrastructure and capacities to address adverse impacts of climate change.

Decisions such as those at the COP 18 and subsequent actions at the global level point to potential compensation mechanism in several directions: direct monetary relief by developed countries to least developed and other vulnerable countries, subsidized transfer of relevant technologies as well as goods and services for improved infrastructure development, launching of risk transfer mechanism on widespread scale (as in crop insurance, property insurance, and other aspects), capacity building (human and technological) to minimize the potential for damages, and possible provision resource channelization from the Green Climate Fund for DRM as an extension of CCA.

A combination of soft laws and binding agreements will evolve over the years to provide better systems of compensation for loss and damage caused by climate change. This debate is likely to engage greater attention as the world continues to confront sustained adverse impacts of climate change as in the cases of extreme events as well as slow onset effects. The concentrations of greenhouse gases will continue to increase; contributions of technical innovations such as carbon sequestration or other mitigation actions are likely to be offset by expanded economic activities with enhanced emissions of greenhouse gases- both production-based and consumption-based. The focus thus needs to be maintained in terms of providing compensation mainly to ensure climate justice and at the same time to build resilient societies so that recurrent adverse events cause least loss and damage. The fundamental links of adverse impacts of climate change and adaptive capacities (among other ingredients explained earlier in Sect. 2.5 deserve closer attention in developing further laws and guidance for actions.

References

FIELD. (2012a). *An option for advancing the UNFCCC negotiations*: *Settling disputes through a conciliation commission*, London: FIELD.

FIELD. (2012b). *Loss and damage caused by climate change*: *Legal strategies for vulnerable countries*, London: FIELD.

Lord, R., Goldberg, S., Rajamani, L., & Brunnee, J. (Eds.). (2012). *Climate change liability: Transnational law and practices: Liability for climate change damage*. Cambridge: Cambridge University Press.

Rao, P. K. (2002). *International environmental law and economics*. Oxford: Blackwell.

Tol, R. S. J., & Verheyen, R. (2004). State responsibility and compensation for climate change damages—a legal and economic assessment. *Energy Policy, 32*, 1109–1130.

Verheyen, R. (2005). *Climate change damage and International Law: Prevention duties and state responsibility*. Boston: Martinus Nijhoff Publishers.

Verheyen, R., & Roderick, P. (2008). *Beyond adaptation: The legal duty to pay compensation for climate change damage*. London: WWF-UK.

Chapter 4
Global Climate Finance

4.1 Introduction

This chapter highlights potential sources of global climate finance mainly for the compensation mechanism in relation to loss and damage, and does not attempt full details of ongoing activities in climate finance generally. Some of the programs such as the Special Climate Change Fund (SCCF) and several others sound big in name and provide meager resources to be able to contribute to tangible results, leave alone constitute even five percent of relief in terms of compensation for loss and damages due to climate change.

Given the potential for substantial financial provisions for climate funds over the next few years under the newly created Green Climate Fund (GCF), an effective integration of CCA and DRR activities is feasible. Policy design and good governance with proper guidance from stakeholders, donors, the UNFCCC, the UNSIDR, and the larger UNDG shall be very useful. The Conference of Parties (COP) for the UNFCCC should explicitly authorize this process.

Funding mechanisms to facilitate relevant and urgent integration are still evolving—potentially via the new GCF. Currently DRR is part of the Least Developed Countries Fund (LDCF) and the SCCF, but not explicitly included in the Adaptation Fund (under the coordination mechanism of the UNFCCC). UN-DAF has taken the lead to provide integration of CCA and DRR. This approach deserves to be extended in all international and national entities dealing with CCA and DRR. This enables synergistic cost-effective linking of actions for effectiveness in the short-run as well as in the long-run. Aligning priorities that allow flexibility over time continues to remain a challenge and further methodological developments are called for. This exercise involves compatibility and balancing among social, economic and environmental objectives of development, and priority setting across sectors over varying time intervals while optimizing resource allocations for maximum productivity and equity.

K. R. Pinninti, *Climate Change Loss and Damage*, SpringerBriefs in Climate Studies, DOI: 10.1007/978-3-642-39564-2_4, © The Author(s) 2014

4.2 Potential Sources of Additional Finance

The Handbook on the OECD-DAC Climate Markers (OECD, September 2011) clarifies that CCA activities are those intended: to reduce the vulnerability of human or natural systems to the impacts of CC and climate-related risks, by maintaining or increasing adaptive capacity and resilience. Inter-governmental organizations (IGOs) and donor nations should adopt this definition and effect integration of CCA and DRR activities. A more balanced focus toward CCA and DRR can emerge and enhanced benefits will ensue.

4.2.1 Green Climate Fund and DRR

In the face of various limitations on public spending, the role of some of the global entities becomes important for developing countries. The newly created GCF should be of help, although it is not necessarily catered to all the needs of DRR.

Although there is no direct role for funding DRR under the new GCF, the effective and efficient route is via an intersection of CCA and DRR projects. It is possible to envisage requirements of larger scale resource outlays when the two streams CCA and DRR are combined but are expected to be more cost-effective than when undertaken in separate sets of activities.

GCF resources allow for technology transfer and concessional funding. This avenue is very relevant for deployment of substantial resources for solar and other renewable energies that belong in the areas of 'No Regret' policies (combining DRR, CCA and CCM) and also enable adoption of SD approaches. When planned appropriately (location, scale and time), these enable resiliency building and infrastructure capacity building to prepare for address adverse impacts of disasters.

Some of the relevant elements of a checklist for implementing international mechanisms include:

Who should pay and who should be paid and how much and what bases for estimation may be used?
Component elements and their valuations for loss and damage;
Limits of direct and indirect liabilities;
Role of insurance and payments for insurance coverage;
Interpreting insurability requirements;
Institutional arrangements for implementation in a cost-effective and time sensitive manner; and,
Incentives for reducing loss and damage in the short-run and in the long-run.

4.3 Other Measures in Climate Change Governance

There are a few additional initiatives on market-based mechanism that are trumpeted under the UNFCCC charter. Market-based and related approaches (UNFCCC 2012e, Document FCCC/TP/2012/4) are summarized very briefly below; these pertain the areas of climate change mitigation and not disaster mitigation, however. Other relevant UNFCCC documents include (UNFCCC, 2012a, b, c, d).

The 'new market mechanisms' (NMM) are being devised in accordance with decision 2/CP.17, paragraph 83, namely "to enhance the cost-effectiveness of, and to promote, mitigation actions". Decision 2/CP.17 states that the NMM is to be guided by the main issues set out in decision 1/CP.16, paragraph 80. These include:

Ensuring voluntary participation of Parties, supported by the promotion of fair and equitable access for all Parties.
Stimulating mitigation across broad segments of the economy.
Safeguarding environmental integrity: approaches "must meet standards that deliver real, permanent, additional and verified mitigation outcomes" and avoid double counting of effort. These standards, as discussed below, remain to be developed in the context of the NMM.
Ensuring a net decrease and/or avoidance of global greenhouse gas emissions.

Possible means to achieve a net decrease and/or avoidance are being attempted with the use of ambitious baselines or targets in the context of the NMM, with the following feature: ensuring good governance and robust market functioning and regulation. Transparency of information and decision-making could be an important element of meeting this objective. Discussions and Party submissions regarding the NMM have suggested two broad approaches to market-based mechanisms: crediting and trading. Given the voluntary participation feature of the NMM it is doubtful if any meaningful results will accrue that can mitigate climate change.

4.3.1 Defining and Selecting Broad Segments of the Economy

The NMM is predicated on the notion of "stimulating mitigation across broad segments of the economy." (decision 2/CP.17, paragraph 80(c)). Though the term "broad segments of the economy" is widely viewed as representing one or more sectors, subsectors, or other groups of emissions sources, Parties have not yet agreed on a precise meaning or definition. A more host country driven approach has been suggested (as compared to Kyoto Protocol mechanisms), with the advantages: better tailoring to national circumstances, better support to national capacity building, and reduced work for international body overseeing the mechanism. Disadvantages may include additional administrative burden on host countries and limited potential to use existing institutions.

4.3.2 Various Approaches

These are the official approaches arising out of some of the COP deliberations and being advanced by the UNFCCC (see for details UNFCCC 2012e at AWG-LCA agenda item 3 (b) (v) *September 2012*).

Various approaches, including opportunities for using markets, to enhance the cost-effectiveness of, and to promote, mitigation actions, considering different circumstances of developed and developing countries, are suggested, taking into account the following:

(a) Ensuring voluntary participation of Parties, supported by the promotion of fair and equitable access for all Parties;
(b) Complementing other means of support for nationally appropriate mitigation actions by developing country Parties;
(c) Stimulating mitigation across broad segments of the economy;
(d) Safeguarding environmental integrity;
(e) Ensuring a net decrease and/or avoidance of global greenhouse gas emissions;
(f) Assisting developed country Parties to meet part of their mitigation targets, while ensuring that the use of such a mechanism or mechanisms is supplemental to domestic mitigation efforts;
(g) Ensuring good governance and robust market functioning and regulation.

If these mitigation actions can result on tangible reduction of greenhouse gas concentrations in the atmosphere we may be somewhere in the region of 2–3 °C warm up, but not enough to control loss and damage due to climate change, however. It is also useful to that a recent study published by the World Bank (2012) highlights the existential threats the world and in particular the vulnerable people in developing countries would face in a 4 °C world, a temperature increase which still can and must be avoided by the international community. However, if mitigation action is not stepped up drastically the world is on the path to dangerous climate change with accelerated occurrence of loss and damage.

4.3.3 Carbon Leakage

Embodied carbon emissions in internationally traded goods have increased both in volume and intensity, and this is the clearest indicator of limited role of mechanisms such as the Kyoto Protocol in addressing the climate change issues or in effecting net decrease in the emissions of greenhouse gases. Effective reduction of carbon emissions will require policy changes in international trade such as the role of carbon taxes and enactment of a new free trade agreement for renewable energy. Covered countries and covered sectors of economic activity need not be limited, although a phase-in period is relevant for each country and sector.

The IPCC (2007) Fourth Assessment Report definition of carbon leakage is too narrow to be useful at the global level, since its reference is simply to climate policy frameworks of countries covered under the Kyoto Protocol. Thus, it has a notion for less than half the total production-based emissions, and that too without any tangible policy guidance to reduce these leakages. The conduit for carbon leakage is international trade.

International trade policies can contribute toward both CCA and DRR. Transformational changes are required in trade agreements to cover relevant goods and services for trade that affect not only reduction of greenhouse gases but also all others to positively contribute to CCA and DRR. The tardiness with which the international trade agreements are making any progress and the mechanisms that fail to keep with relevant advancements in science and technology are least encouraging. The lofty objectives and goals of the WTO charter are yet to be fully realized. Let us recall that the Preamble to the Marrakesh Agreement Establishing the WTO states:

> The Parties to this Agreement [recognize] that their relations in the field of trade and economic endeavour should be conducted with a view to raising standards of living, ensuring full employment and a large and steadily growing volume of real income and effective demand, and expanding the production of and trade in goods and services, while allowing for the optimal use of the world's resources in accordance with the objective of sustainable development, seeking both to protect and preserve the environment and to enhance the means for doing so in a manner consistent with their respective needs and concerns at different levels of economic development ...

An effective operationalization of these objectives can reduce potential L and D due to climate change, and also reduce adverse impacts of trade and on climate change.

References

IPCC. (2007). *The fourth assessment report: Climate change.* Geneva: IPCC.

OECD. (2011). *The handbook on the OECD-DAC climate markers.* Paris: OECD.

UNFCCC. (2012a). *Current knowledge on relevant methodologies and data requirements as well as lessons learned and gaps identified at different levels, in assessing the risk of loss and damage associated with the adverse effects of climate change.* UNFCCC Technical Report FCCC/TP/2012/1, Bonn.

UNFCCC. (2012b). *A literature review on the topics in the context of thematic area 2 of the work programme on loss and damage: a range of approaches to address loss and damage associated with the adverse effects of climate change.* UNFCCC Paper FCCC/SBI/2012/INF.14, Bonn.

UNFCCC. (2012c). *Report on the regional expert meetings on a range of approaches to address loss and damage associated with the adverse effects of climate change, including impacts related to extreme weather events and slow onset events.* UNFCCC Report FCCC/SBI/2012/ 29, Bonn.

UNFCCC. (2012d). *Report on the expert meeting on assessing the risk of loss and damage associated with the adverse effects of climate change.* UNFCCC Document FCCC/SBI/2012/ INF.3, Bonn.

UNFCCC. (2012e). *Various approaches, including opportunities for using markets, to enhance the cost-effectiveness of, and to promote, mitigation actions, bearing in mind different circumstances of developed and developing countries.* UNFCCC Technical paper FCCC/TP/ 2012/4, Bonn.

World Bank. (2012). *Turn down the heat: Why a 4 degree C warmer world must be avoided.* Washington: World Bank.

Chapter 5
New Frameworks for Financing and Governance of Loss and Damage

5.1 Roles of Historic Contributions and Responsibilities

Financing and cost-sharing for compensation mechanisms largely centers around developed countries for their historic contributions even since industrialization started. However, using the UNFCCC charter (see also Muller et al. 2009), especially Article 3.1 which refers to 'contribution' 'respective capabilities' and 'common but differentiated responsibilities', suggests considerable need for precise application of the norms with a good degree of empiricism. This is not a part of this Monograph. Among the estimates of burden sharing offered for financing CCA an estimate of about $100 billion per year has been suggested for supporting CCA (Dellink et al. 2009); it is also suggested that a combination of ability to pay and historical contributions might be relevant, and the per capita burden for compensation in this context among industrial countries ranges $43–82 per year. Suffice it state that if the order of magnitude for compensation for L and D (beyond CCA financing) is about $120 per year the per capita may not be high for the developed countries. These magnitudes and higher levels of support could involve co-financing of insurance coverage and other risk spreading mechanism, concessional transfer of technology that bring in economies of scale, and related measures that promote the businesses in the developed countries as well. Thus a set of win–win strategies are relevant in this context.

5.2 Effective Integration of CCA and DRR

The fullest integration of CCA and DRR should remain a long term goal, in the interests of efficient and adaptive resource allocation and governance. In the medium term and short-term, a series of progressive measures can be undertaken by the national governments. DRR activities can largely accommodate CCA activities and more. Since the design and implementation of the fullest extent of relevant CCA in any system is infeasible, there always remains residual damage

K. R. Pinninti, *Climate Change Loss and Damage*, SpringerBriefs in Climate Studies, DOI: 10.1007/978-3-642-39564-2_5, © The Author(s) 2014

(RD) in CCA. National governments need to balance the resources for CCA and for managing RD. The roles of DRR and DRM directly belong here. These assessments are necessary under the NAPs and NAPAs governed by the UNFCCC systems.

As the last major principle we list the Precautionary Principle, captured in Principle 15 of Rio Declaration: "In order to protect the environment, the precautionary approach shall be widely applied by States according to their capabilities. Where there are threats of serious or irreversible damage, lack of full scientific certainty shall not be used as a reason for postponing cost-effective measures to prevent environmental degradation." This principle is based on the idea that uncertainty (e.g. with regard to any environmental problems such as biodiversity loss which has biological, ecological as well as economic implications) should be treated with a measure of safeguard—in fact the precautionary principle reflects a "better safe than sorry" principle, "risk averse" or "no regrets" policy (Rao 2000). This principle has been critical when debating climate change, genetically modified organisms, and other environmental risks. It can be formulated as either 'states should take action to protect the environment even in case of scientific uncertainty' (as per the Principle cited above) or as 'states should refrain from action potentially damaging the environment even in case of scientific uncertainty' (as in the Cartagena Protocol on Biosafety to the Convention on Biological Diversity, Art. 10, para. 6). Although the principle might be difficult to apply in a policy context, since it only recommends the direction (e.g. a reduction of GHG emissions) of a policy action rather than its corresponding magnitude (e.g. the amount of reduction necessary), it renders an important dimension of SD; it implies current commitment to safeguard against the likelihood of future occurrence of adverse impacts, being related to the principle of intergenerational equity (Rao 2000).

Integrating CCA, DRR, and SD is the way to go. However, substantial learning is due in these integration efforts at various levels and spheres of activity. Considerable analyses and in-depth knowledge on relevant linkages are available. The IPCC Special Reports on Extreme Events (SREX Report, IPCC 2012), and on Renewable Energy (SREN Report, IPCC 2012), and several other recent papers and reports are very relevant in devising relevant strategies. However, *relevant actions in any follow up are lagging far behind. Public policy frameworks and actions missing on the scientific knowledge are similar to living in the past with little preparedness for the future; this constitutes an inertia and lethargy-based approach toward living for the collective doom of all.*

5.3 Roadmap for Effective Governance

Poverty remains a common drag on all the elements of common frameworks for CCA, DRR, SD and is an integral part of the original definition of SD as per the WCED 1987 Report. However, this underlying factor is not yet understood or deliberately neglected in many policy designs which claim they are meeting some of the SD imperatives. Similarly, many poverty reduction strategies (for example the Poverty Reduction Strategy Papers of the IMF/World Bank) do not yet see the urgent need for integration of DRR factors. *The emerging post-2015 frameworks of Development Agenda and of Sustainable Development Goals should be able address these aspects, and also realize that disasters contribute to the worst forms of poverty: chronic poverty and ultra-poverty. These constitute a permanent severe drag on a significant section of population, and for the rest of the socio-economic system. Target setting on these dire features in conjunction with goal setting for an effective integration of CCA, DRR and SD will be cost-effective both in the short-run and in the long-run.*

In terms of planning and decision-making approaches for this integration, nations and international entities will do better with the adoption of a reasonable sequencing of win–win–win combinations of projects that avail 'no regrets' and 'low regrets' approaches: reduce poverty with CCA activities combined with DRR, seek priorities with reduction of children and women, seek a risk-balanced portfolio of projects that simultaneously incorporates the above, plus do not miss out on the missing governance aspects. *Effective and efficient delivery of results with active participation of stakeholders remains a desirable approach.*

Existing research that explores the linkages between poverty reduction and hazard risk reduction has mainly focused on assessing poverty outcomes of large-scale catastrophic hazards. While such events have extreme impacts on poor populations, their infrequence makes it very difficult to establish a relationship with poverty trends over time, except at the macro-level. In contrast, there is a large number of frequently occurring but highly localized events, such as land-slides, flash floods, fires and storms that may represent a significant and unreported source of losses and disruption to livelihoods for marginal rural and urban pop-ulations. These circumstances may therefore have a crucial interactive relationship with poverty patterns and trends. There is now recognition that it will become necessary to focus more attention on the impacts of highly localized, low intensity hazards on poverty.

The fungibility of aid and of project assets is rather widely seen as a major problem that afflicts DRR and DRM activities, even more than in other sectors. For example, repair works of one season may not have a lasting for the next, and accountability mechanisms for construction and repair works are very weak in most countries. The capabilities of organizations for conducting monitoring and evaluation (almost on an ongoing basis) remain limited, and are usually expensive. *The roles of local community organizations can be effectively tapped to enhance quality of works, reduce corruption, and reduce total costs of governance on a recurrent basis. Initial capacity development including formation and calibration of effective community organizations paves the way for the creation of cost-effective and durable infrastructure (Rao 2013).*

Resource requirements to cover loss and damage among vulnerable developing countries are in terms of concessional technology transfer, subsidized insurance mechanism, funds for DRR, resources for transformative CCA (much beyond minimalist incremental work).

Article 3.1 of the UNFCCC stipulates that Parties "should protect the climate system ...in accordance with their common but differentiated responsibilities and respective capabilities".

There are several proposals in operationalizing the norms of CBDR and respective capabilities (see for example, Muller and Mahadeva 2013; and Page 2008). These explore possible assessments of ability to pay, contribution to climate change, and the role of beneficiary pays criteria. None of the are without their own shortcomings, and it is not proposed to review the details in this Monograph.

In the UNFCCC framework, "Various approaches" need to be replaced with effective and cost-minimizing cooperative solutions in the short-run and in the long-run. Cost-effective adverse-impact reducing mechanisms deserve consider-able further analysis, taking fully into account the role, not just of neo-classical economics that dominates the economic approaches, but availing the applications of the economics of Transaction Costs and New Institutional Economics.

An earlier report (UNFCCC 2007) suggested the role of the following mechanisms:

Risk management and risk reduction strategies, including risk sharing and transfer mechanisms such as insurance;

Disaster reduction strategies and means to address loss and damage associated with climate change impacts in developing countries that are particularly vulnerable to the adverse effects of climate change;

Provision of financial and other incentives for, scaling up of the development and transfer of technology to developing country Parties in order to promote access to affordable environmentally sound technologies;

Ways to accelerate deployment, diffusion and transfer of affordable environmentally sound technologies; and,

Enhanced action on the provision of financial resources and investment to support action on mitigation and adaptation and technology cooperation.

Clearly, there has been a good deal of recognition of the relevant implementable activities but actions are lagging far behind. Losing time is not an option for the developed world because of the potential to compensate for loss and damage increases over time in direct non-linear proportion to the delay factors. For the vulnerable, significant and irreversible loss and damage occurs with greater scale and rapidity over time. Thus there is an urgent need for pragmatic time-bound actionable framework.

Managing risks for development remains a key determinant of building resilient economies and resilient societies around the world. A better understanding of roles of short-term resources and provision of long-term incentives for risk reduction is critical to the advancement of such development framework. Recurrent nature of events and building resilience are to be addressed with better involvement of stakeholders in various stages of operations: planning, implementation, monitoring and evaluation as reforms.

The role of the ICTs needs to be expanded in a significant transformative approach that contributes toward enhanced CCG, and DRR. There are synergistic win–win strategies for the private sector and the governmental entities in this expanded framework. The newly adopted October 2010 Resolution at Guadalajara "The role of Telecommunications/Information and Communication Technologies on Climate Change and the Protection of the Environment" identifies the need to assist developing countries to use information and communication technologies (ICTs) to tackle climate change and committed the International Telecommunications Union (ITU) to work with other stakeholders to develop tools to support this aim.

References

Dellink, R., Elzen M den, Aiking H., Bergsma E., Berkhout F., Dekker T., Gupta, J. (2009). Sharing the burden of financing adaptation to climate change. *Global Environmental Change. 19*, 411–421

IPCC. (2012). *Managing the risks of Extreme Events and Disasters to Advance ClimateChange Adaptation.* from http://ipcc-g2.gov/SREX/images/uploads/SREXSPMbrochure_FINAL.pdf .

Mueller, B., & Mahadeva, L. (2013). *The Oxford Approach: Operationalising the UNFCCC Principle of Respective Capabilities, Report EV 58.* Oxford: The Oxford Institute for Energy Studies.

Mueller, B., Hohne, N., & Ellermann, C. (2009). Differentiating (historic) responsibilities for climate change. *Climate Policy, 9*, 593–611.

Page, E. A. (2008). Distributing the burdens of climate change. *Environmental Politics, 17*, 556–575.

Rao, P. K. (2013). *Governance of Disaster Reduction.* Working Paper, Geneva: UNISDR.

Rao, P. K. (2000). *Sustainable Development: Economics and Policy.* Oxford: Blackwell.

UNFCCC. (2007). *Report of the Conference of the Parties on its 13th Session*, Bali, December, 2007, Report FCCC/CP/2007/6/Add.1, Bonn: UNFCCC.

Chapter 6
Concluding Observations

The 'ultimate objective' of the UNFCCC, "to achieve stabilization of greenhouse gas concentrations in the atmosphere at a level that would prevent dangerous anthropogenic interference with the climate system" is at risk.

It is urgent and important that new international agreements recognize the dangerous impact that climate change is already having on at-risk communities and ecosystems, and enact measures to both reduce risks of and respond effectively to the adverse effects. The UNFCCC should provide the framework for addressing climate-related loss and damage and scaling up required actions, as outlined before. L and D, with its different dimensions, will need to be a core part of the new global agreement on climate change. *New agreement(s) needs to be considered seriously in the Durban Platform process towards developing a 2015 agreement and ensure coherence with various post-2015 agendas and frameworks in progress: those of the HFA, MDGs, Sustainable Development Goals and Development Agenda.*

Besides, the emerging post-2015 framework for sustainable development goals needs to incorporate the potential targets for contributions from developed countries towards compensation for climate change L and D (see also Hyvarinen 2012).

The scale of immediate resource needs toward various elements that constitute compensation for loss and damage may be at least of the order of about US$ 112 Billion per annum. This number is a tenth of the total loss and damage estimate offered by DARA International for the year 2010 (see also Sect. 2.4). *Issues that remain to be resolved are how much of this may be apportioned from the GCF and how much can be contribute by other mechanisms (including public–private partnership).* However, a 4 degrees C global average warm up possibility can lead to disproportionately higher estimates of L and D.

The decision from the Durban conference stated the "need to explore a range of possible approaches and potential mechanisms, including an international mechanism, to address loss and damage". Multiple entry points to reduce loss and damage, contained in the Cancún Agreements, have also been recognized; these include: climate risk insurance facility, options for risk management and reduction, risk sharing and transfer mechanisms such as insurance, resilience building, and approaches for addressing rehabilitation measures associated with slow onset events.

K. R. Pinninti, *Climate Change Loss and Damage*, SpringerBriefs in Climate Studies, DOI: 10.1007/978-3-642-39564-2_6, © The Author(s) 2014

In terms of potential magnitudes for compensation for L and D, total restitution may not be a viable option (FIELD 2012), but a pooled CC Risk Facility may be a desirable with an element of urgency both in its formation, efficient administration, and delivery of resources in a time-sensitive cost effective manner.

Climate-related loss and damage is too big an issue to be resolved only by an environmental agreement such as the UNFCCC. Issues such as human mobility or loss of territory will require consideration from other global bodies, including the UN Security Council and the UN High Commission on Human Rights Council (see also Action Aid International 2012).

Disaster losses and their 'normalized trends' even in developed countries such as the USA are assessed as unsustainable under the current technical and institutional configurations. The situation concerning most of the developing and especially vulnerable countries remains too serious and warrants pragmatic measures that should cut across organizational boundaries in the national and international frameworks. It is neither the sole monopoly nor responsibility of a single entity such as the UNFCCC to cater to the relevant needs. This applies also to relatively smaller organizations such as the UNISDR. This should not, however, imply a suggestion toward abdication of relative responsibilities of each of such organizations. *A high-level intergovernmental coordination mechanism that encompasses such entities as the GCF, GEF, UNFCCC, UNISDR, WTO, and a few others is urgently called for.*

Role of supporting services such as early warning systems and climate services has not been fully utilized, as seen, for example, in the limited adoption and resource provision in various NAPAs. UNFCCC (2012) Report in its end data is illustrative of the types of disaster impact reducing plans and actions various countries implemented or planning to implement. These appear attuned to their local sense of priorities and a more comprehensive approach and strategy that is owned by individual countries is called for. The rough estimates of resource requirements for compensation toward loss and damage seem to be of the order of the current levels of targets of official development assistance (ODA) from developed countries to others.

The role of ICTs and relevant infrastructure remains critical both in cost-effectiveness (as demonstrated various assessments of the economics of EWS) and in time-sensitive effective provision of goods and services. Appendix to this Monograph summarizes some of the key features that pertain in the context of minimization of loss and damage due to adverse events.

There are international and national responsibilities—separately and jointly for reducing: (a) factors contributing to climate change—induced adverse events (both extreme events and slow onset events); and (b) impacts of these events so that all natural hazards need not result in disasters. These assertions have implications for the allocation of resources and governance a various levels. Prevention activities are applicable at all levels and constitute cost-effective measures, and so are effective governance mechanisms at various levels. Compensation for loss and damage is a resultant some of these actions. In a dynamic framework for decision-making asset preservation and maintenance remain fundamental requirements—for insurability, risk minimization, and risk compensation. National and international frameworks (including the post-2015 frameworks of Sustainable Development Goals) need to focus on resilience building in vulnerable nations and offer broader framework for mitigation of loss and damage due to climate change. Simultaneously the UNFCCC charter can devise pragmatic comprehensive principles for adoption in this context, noting that the adverse impacts of climate change tend to exceed the gross domestic product—as a global average on annual basis. We need to avoid one step forward and two steps backward in the formulation of policies and devising institutions of governance.

It is apt to quote Nobel Laureate Ostrom (2010) who focused on the end product of minimizing impacts of adverse effects due to climate change at multiple scales with the roles of stakeholders at all levels:

> Climate change is a global collective-action problem since all of us face the likelihood of extremely adverse outcomes that could be reduced if many participants take expensive actions...... instead of focusing only on global efforts (which are indeed a necessary part of the long-term solution), it is better to encourage polycentric efforts to reduce the risks associated with the emission of greenhouse gases.

Finally, L and D compensation approaches and mechanisms should adopt cost-effective solutions for reducing L and D on a sustainable basis, while recognizing short-term priorities.

References

Action Aid International, CARE International, German Watch, & WWF. (2012). *Into Unknown Territory: The Limits of Adaptation and Reality of Loss and Damage from Climate Impacts*, www.actionaid.org

FIELD. (2012). *Loss and damage caused by climate change: Legal strategies for vulnerable countries*. London: FIELD.

Hyvarinen, J. (2012). *Loss and damage caused by climate change: Legal strategies for vulnerable countries*. London: FIELD.

Ostrom, E. (2010). Polycentric systems for coping with collective action and global environmental change. *Global Environmental Change, 20*, 550–557.

UNFCCC. (2012). *A literature review on the topics in the context of thematic area 2 of the work programme on loss and damage: A range of approaches to address loss and damage associated with the adverse effects of climate change*. UNFCCC Paper FCCC/SBI/2012/INF.14, Bonn.

Appendix
Role of the ICTs and ITU in CCA and DRR

The following is largely based on Bueti and Faulkner (2012) regarding potential contributions of ICTS and the International Telecommunications Union (ITU) in the context of devising cost-effective and result-oriented disaster management strategies that can contribute toward mitigation of disasters and constitute elements of compensation mechanisms for L and D.

The linkages between information and communications technologies (ICTs) and climate change adaptation are significant. ICTs help advance weather forecasting and climate monitoring, and also disseminate information to large sections of the society, for example via mobile phones or 'reverse calling' when local authorities as in some municipalities in the USA provide for such emergency alerts for the civic community. This can help address major adaptation risks such as food and water shortages through providing early warning systems and better monitoring of relevant focus features.

The ICTs including remote sensing and geographic information systems have expanded the possibilities for risk assessment of multiple hazards and enabled the development of various scenarios and contingency plans. Risk analysis includes: risk maps, hazard maps, and scenario maps, and ex post assessment based on GIS. Risk analysis is thus a key component in developing a DRR strategy by establishing the links between exposure to hazards, level of vulnerabilities and the capacity to address the hazards.

Over 7,000 natural disasters occurred during 1980 to 2005 worldwide in which millions of lives were lost. Ninety percent of these disasters were caused by weather and water related events such as floods, cyclones and droughts. Access to information and increasing knowledge among policymakers and the general population is part of 'capacity building'. In terms of the telecommunications networks 'capacity building' has an additional meaning which is the expansion of telecommunications networks to serve greater numbers of the population. Adequate telecommunication networks are essential in ensuring that communications reach people and the appropriate relief organizations.

An example of how ICTs can help in reaching people in remote areas is the 'Green Power for Mobiles' initiative which is pioneering alternative power sources such as solar and wind for mobile base stations to serve the one billion people without access to grid electricity. The benefits of such initiatives include reaching

K. R. Pinninti, *Climate Change Loss and Damage*, SpringerBriefs in Climate Studies, DOI: 10.1007/978-3-642-39564-2, © The Author(s) 2014

more people with climate related information and alerts, and to improve coverage of environmental monitoring systems with greater reliability.

Complex emergencies need external intervention in resource-poor countries where data and communication facilities are scarce. Decision-making is often delayed due to lack of information. The effectiveness of humanitarian interventions and the ability to protect livelihoods from the impacts of hazards depends on the timeliness and appropriateness of responses. In order to minimize loss and damage it is important that infrastructure is upgraded with priority for ICTs.

The role of the ITU includes (Bueti and Faulkner 2012):

ITU provides assistance to governments to build appropriate institutions for disaster risk reduction; develops international standards; provides assistance to countries in incorporating resilient features in telecommunications infrastructure; helps countries to develop policy and legal frameworks by providing inputs into policy formulation, and legislative and regulatory drafting for countries; helps countries with regard to their vulnerability by providing assistance in reducing and eliminating vulnerabilities in telecommunications infrastructure; assists Member States in designing and incorporating telecommunications/ICT into national adaptation plans; implements early warning systems in countries where there is a high incidence of disasters; designs national emergency telecommunications plans that include Standard Operating Procedures that are now in use in many countries; produces guidelines, toolkits and other publications that are in use by countries for disaster risk reduction.

Reference

Bueti, C. and Faulkner, D. (2012). ICTs as a Key Technology to Help Countries Adapt to the Effects of Climate Change, Geneva: International Telecommunications Union.